T0200239

Owing to this struggle for life, any variation, however slight and from whatever cause proceeding, if it be in any degree profitable to an individual of any species, in its infinitely complex relations to other organic beings and to external nature, will tend to the preservation of that individual, and will generally be inherited by its offspring. The offspring, also, will thus have a better chance of surviving, for, of the many individuals of any species which are periodically born, but a small number can survive. I have called this principle, by which each slight variation, if useful, is preserved, by the term of Natural Selection.

Charles Darwin, *The Origin of Species* (1859)

Series 117

This is a Ladybird Expert book, one of a series of titles for an adult readership. Written by some of the leading lights and outstanding communicators in their fields and published by one of the most trusted and well-loved names in books, the Ladybird Expert series provides clear, accessible and authoritative introductions, informed by expert opinion, to key subjects drawn from science, history and culture.

The Publisher would like to thank the following for the illustrative references for this book:
Cover and page 9 © Henry Guttmann/Hulton Archive/Getty Images; page 17 © Bettmann/Getty Images.

Every effort has been made to ensure images are correctly attributed, however if any omission or error has been made please notify the Publisher for correction in future editions.

MICHAEL JOSEPH

UK | USA | Canada | Ireland | Australia
India | New Zealand | South Africa

Michael Joseph is part of the Penguin Random House group of companies
whose addresses can be found at global.penguinrandomhouse.com

Penguin
Random House
UK

First published 2017
002

Text copyright © Steve Jones, 2017

All images copyright © Ladybird Books Ltd, 2017

The moral right of the author has been asserted

Printed in Italy by L.E.G.O. S.p.A.

A CIP catalogue record for this book is available from the British Library
ISBN: 978–0–718–18628–9

www.greenpenguin.co.uk

Evolution

Steve Jones

**with illustrations by
Rowan Clifford**

Ladybird Books Ltd, London

A sketch of a sketch

Nobody can speak a language without understanding how it fits together. Evolution is the grammar of biology. It unites the study of plants, animals and people into a single science. Without it, the subject would be a list of disconnected facts, as it was until 1859, when Charles Darwin published *The Origin of Species*.

His ideas led to a revolution in science and in humankind's view of itself. Although creationists still believed that life had sprung into being around 6,000 years ago, and although some biologists had speculated that living organisms could change, they had no real idea how or why, and few facts to back up their ideas. Darwin's theory, by comparison, was almost brutal in its simplicity. It turns on 'descent with modification': the accumulation of errors over the generations. Diversity, its raw material, is refined in the furnace of natural selection: inherited differences in reproductive success in the face of a struggle for existence.

Darwin accumulated so much evidence that he might never have published had he not received a letter pre-empting his idea from the naturalist Alfred Russel Wallace. Darwin rushed out a 'Sketch' of his theory. He calls *The Origin* 'one long argument' and, like a legal case, it moves from the obvious to the almost unthinkable: from cattle breeding to the claim that 'light will be cast on man and his origins'. That statement has proved triumphantly correct.

If *The Origin* is a 'Sketch' at 170,000 words, the Ladybird Expert book *Evolution* is sketchier at a mere 6,000. Even so, I hope that it reveals the bare bones of modern evolution.

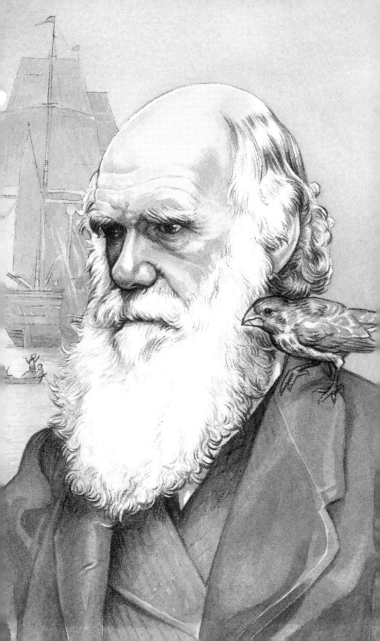

A hippo in the Himalayas

Hippopotami are rare on Everest today. However, they are there in spirit, for their fossils and those of their ancestors are scattered across the peaks. Some of their descendants still roam African swamps, but many more have changed their lifestyles, and sport in the deeps instead.

Genesis says that 'God created great whales', but Darwin dared to criticize that idea. Science has now demolished it, and the whales show how.

Evolution is a series of successful mistakes: a factory for almost impossible things. Millions deny the power of natural selection to generate eyes or whales, but they are wrong.

Fossil whales are in the Himalayas not because they moved there, but because the Earth did. Long ago those mountains were beneath the ocean, but as continents collided they were thrust into the skies. The fossils show a seamless transition, from hippo-like creatures 55 million years ago to others that, like polar bears, swam in the sea with four legs, to versions rather like seals, to animals that lost their forelegs until at last, around 25 million years ago, fins, blowholes and modern whales made their appearance. Divine assistance was not needed.

Those great creatures still share some characteristics with hippos – they squeak to communicate, lack hair, and have testes inside the body. More important, the message written in whale DNA is more similar to that of modern hippos than to that of elephants or even seals.

Darwin on HMS *Beagle* saw plenty of whales, but he knew nothing of their past. Now we do, and in more detail than for any other mammal.

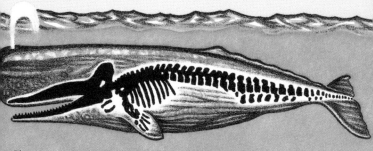
Physeter macrocephalus, the sperm whale

Dorudon

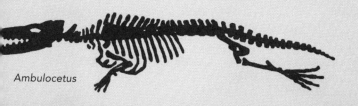
Ambulocetus

Pakicetus

Evolution makes sense

The Origin points out that language, like life, changes over the generations as mutations – shifts – in the way that people speak build up. I hear that every day among my students, most of whom now affect a half-cockney accent that must horrify their grandparents. Language also illustrates the power of selection as it winnows out failures and rewards messages that make sense.

A pig stands outside its sty. The challenge is to get the beast inside in the fewest possible steps, by changing each of the three letters in its name one at a time. Given the number of letters in the alphabet, to do the job by simple random error – mutation – would need an average of 17,576 tries ($26 \times 26 \times 26$) before the goal is reached. Impose a simple rule, that only combinations that represent English words are allowed to survive, and it can be done in six – pig, wig, wag, way, say, sty.

To make by mutation alone a real pig from its ancestor – even a wild boar, let alone the shared progenitor of pigs, hippos and whales – would take more tries than there are stars in the universe. Natural selection got the boar on to the farm in a few thousand generations, and from the ancient mammal in a few million. From pigs to those who domesticated them, the history of life depends on making sense.

pig

wig

wag

way

say

sty

Eat your greens

Domestic animals make a powerful case that flesh can change as farmers breed from animals that make more milk, more wool or more eggs; but the plant world gives even better evidence of the power of selection.

It works fast. A wild cabbage that grows on sea-cliffs has seen dramatic shifts in personality. In Imperial Rome gardeners had already produced a variety with thicker leaves – kale. Around the time of Christ, a version with a tight head made of immature leaves – cabbage – joined it, while the Germans began to breed from plants with stout stems, now known as kohlrabi. The Renaissance saw the birth of the cauliflower, selected for large flowers, and, in Italy, of its close relative broccoli. In the eighteenth century Belgian farmers began to select varieties with tight buds, and the Brussels sprout joined the other family members on the plate. There has been change even in the recent past, with white, red and black cabbage, plain and sprouting broccoli, and normal and flower sprouts (which emerged only ten years ago).

If cabbages and the rest descend from the same ancestor through human preference over a few hundred or a few dozen generations, then could not Nature do much more to form new plants and animals with so many years, and so many individuals, in her armoury?

Barking at the moon

Darwin loved dogs, both as pets and as proof of the efficacy of selection. As he pointed out, they, like many domestic animals, emerged through changes in the brain as much as the body.

In 1959 the Russian biologist Dmitri Belyaev set out to breed silver foxes, their fur much used for winter hats. At first the animals were terrified by humans and were almost impossible to deal with. He began to select those most ready to allow him to approach their cages without going into a frenzy and bred only from the tamest 5 per cent. Within a few generations there was a remarkable change. The animals became calm, they began to bark and wag their tails and, like puppies, loved to be handled. They looked different, too; not black with a few silver hairs, but much paler, with – like many cattle, horses, cats and dogs – a large white spot on their face or chest (they also like marshmallows). They now sell for thousands of dollars as pets.

The machinery that makes the black pigment melanin is related to that involved in the mood-stabilizing hormone serotonin, so that the colour shift and behavioural change are symptoms of the same response to evolutionary pressure. Inside many wild creatures is a tame one struggling to get out. With a simple trick, breeders can set it free.

Accounting for taste

Soon after his election in 1988 the first President Bush said: 'I'm President of the United States and I'm not going to eat any more broccoli!' He is joined in his dislike by millions. But why?

His statement is a hint of the enormous diversity hidden in every population. Broccoli-haters have inherited a single DNA change in the 3,000 million-letter message that makes them reject bitter food (they are also less likely to smoke or drink). It sits within one of hundreds of genes that determines individual preference for a sweet, salty or fatty diet and is just one of millions of individual differences scattered across the genetic message. They involve chemical changes in the letters of the double helix, deletion or expansion of sections of the molecule, reversals in the order of sections of its letters, and more.

Sex reshuffles those DNA cards each generation. As a result, every man and woman – or fox, or dog, or whale – alive today is different from every other member of their species that has lived, or ever will live. Inherited differences affect our appearance, our behaviour, our health and our evolution and, most of all, give each of us a unique set of strategies when playing biological poker against the ever-changing world outside in a struggle for existence in which a losing hand means oblivion.

An unhappy accident

When I was eight, in 1952, I suffered from a minor illness which ten years earlier might have killed me. A cut became infected, and my hand swelled up mightily. A quick jab with a scalpel and a couple of injections solved the problem, for penicillin was then in its heyday.

It did not last. The drug was introduced in 1944 by Alexander Fleming, but within a decade many bacteria were able to resist it and in modern times it has become almost useless. New inherited changes, mutations, are to blame. These arise at a rate per gene of just one in a million per generation, but bacteria come in billions, which means that in the end mutations are inevitable. With their help, the bacteria keep drugs out, hide them away where they cannot harm them, or break them down, depending on what mutations turn up.

Penicillin has had many successors but resistance always arises, sometimes within a few months. Part of the problem comes from the reckless use of antibiotics on farms and part from the ignorance of doctors who prescribe them for flu and other problems caused by viruses rather than bacteria. Now, medicine has almost none left in its armoury and almost nothing seems to be around the corner.

I still have a scar on my right hand as a reminder of the dangers of mutation. The power of its twin, natural selection, means that the next generation might not be so lucky.

A fresh start

All mutations represent new opportunities, favourable or otherwise, for those who inherit them. Fish tell the story. The last Ice Age ended 10,000 years ago. Most sticklebacks live in the ocean, but, all around the northern hemisphere, as the glaciers melted some ventured upstream into the new rivers and lakes. At once, mutation and selection swung into action.

Life at sea is hard, with many predators. The marine kinds (*above*) have evolved, as a result, thick armour and fearsome spines. Once in fresh water, existence was easier, for most of their enemies could not follow. Spines and shields are expensive and were soon discarded, so that today's freshwater sticklebacks (*below*) are much less formidable than their ancestors. Selection picked up mutations that blocked the action of just two genes, one that codes for a thick skin, and another active in the lower back, where it caused spines to grow.

It did so, quite independently, again and again across the northern world as new rivers opened up. In one Alaskan lake, the marine form invaded just a decade ago, and already has begun to shed its armour. As global warming makes the ice fall back, evolution is still hard at work. There are far fewer sticklebacks than there are bacteria, but for them, as for all other creatures, mutation will, sooner or later, give them the tools for change.

Wings, fins and echoes

Different species, as much as different populations of the same species, often come up with the same evolutionary solution, although they may reach it in different ways.

Birds and bats both have wings, but bats use a membrane stretched between an arm and five elongated fingers while birds just have a reduced arm, with added feathers. Whales and sharks each have a tail fin, but the mammal's is horizontal and that of the fish vertical; whales flex their bodies and gallop through the sea, while sharks just wag their rear ends. In each, selection picked up what mutation offered to rise to an environmental challenge.

Other patterns are less obvious. Whales, bats and a few birds each bounce waves off nearby objects to hunt and to navigate. Bats emit high-pitched squeaks which work up to ten metres away, while certain whales send out clicks a thousand times more powerful that, underwater, reach thirty times further. The cave-dwelling oilbird of South America uses high-pitched clicks to home in on its nest from a few metres. Each creature also has changes in their ears that improve their ability to pick up faint echoes. In the same way, eyes have evolved, quite independently, dozens of times while Australia has creatures that look like mice, dogs, moles and flying squirrels which are, in fact, entirely unrelated to them. In evolution, parallel lines often converge.

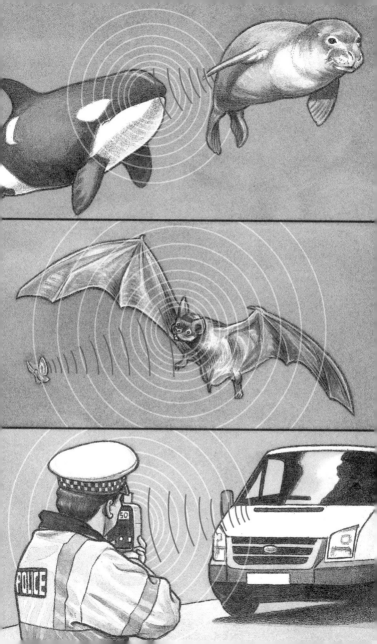

Hearing from the past

Powerful as natural selection may be, it does not plan ahead. Quite often, complex organs are cobbled together with bits and pieces of other structures. If life was designed by an engineer, he would soon lose his job.

The ear is a classic of such opportunism. It has outer, middle and inner segments, which transform sound waves into nerve impulses. Each segment has a separate history. The outermost focuses sound waves onto the eardrum, which vibrates in sympathy. On its inner surface rests the first of three interlocking levers which pass that signal to a second membrane that seals a fluid-filled bony spiral. That is lined with flexible sensors sensitive to waves of different frequency. They bend in response to movements of the liquid and convert them into electrical signals to the brain.

The outer ear is part of the skin, while the middle ear descends from the gills of fish and the jaws of reptiles, as manifest both in fossils and in the early embryo, in which gill-like structures develop into its levers. The inner ear traces back to the fish lateral line – a set of pressure-sensitive receptors used to navigate and to hunt by picking up waves in the water. Ears, and many other structures, have evolved by tinkering rather than industrial design; they do their job as well as they need to, but no better.

An inspector calls

Any machine, designed or evolved, in a factory or in a living organism, needs careful supervision. Natural selection spends most of its time acting as a quality superintendent, always alert to throw out faulty products. Sometimes, though, a structure loses its job. Then, the inspector loses interest.

At once, errors build up. Whales do not have legs, but retain bony vestiges of what they once were. Cave fish have no eyes, and even their genes for light-sensitive pigments have withered away.

We have just the same history. Birds and reptiles have four such pigments, one of which picks up short-wave ultraviolet radiation, which we cannot perceive. As a result, males and females of the same species, which may look identical to us, sometimes appear quite different to the birds themselves, with the male alone bearing bright marks (*top*). Like many mammals, we have receptors for three colours, red, blue and green (*middle*), and millions of colour-blind people have just two and cannot distinguish green from red (*bottom*). We may have lost one of the four originals when, in the days of the dinosaurs, our early ancestors came out mainly at night.

The decay in vision is as nothing compared to that in taste and smell. Around 800 genes are involved. In mice and dogs most of them work, but in humans hundreds have been rendered useless by mutation. Why they stay there we do not know. Evolution, like all sciences, is full of unanswered questions.

The joys of sex

Most of the beautiful sounds, sights and smells of nature are screams of sexual frustration. Flowers, birdsong, mandrills' bottoms and faces (*above*) and red deer's antlers (*below*) are all statements of masculine lust and feminine response to it.

Natural selection is an examination with two papers. The first is simple, for it just involves staying alive. The second is harder, for it is about sex. Males and females must answer different questions, and the former have a higher failure rate. Darwin once said that 'the sight of a feather in a peacock's tail makes me sick'. At first he could not see why males had huge bright tails that did not help them fly, but females did not. He soon realized that it came from sexual differences. Females are limited in how many progeny they can have because of the costs of pregnancy and child-care. Males, in contrast, could – if they can attract enough partners – have vast numbers, which means that many of their fellows are doomed to disappointment.

Males hence struggle to succeed against their sexual competitors while females try to select the finest possible mate. Males flaunt their quality by fighting or with ornaments that only the best can afford. Even colours cost a lot. Red, yellow and orange are often based on complex chemicals that are hard to make, so that hungry, diseased or stressed swains find it more difficult to maintain them and so do less well in the mating game. There is, alas, not much romance in the world of nature.

Bullies, cheats and liars

There is, however, plenty of bullying. The male partner in land snails shoots a 'love dart' into its mate, not as a statement of affection but as a weapon coated with hormones that force the female to accept his sperm.

Other males prefer to lie. Sneaky (and weedy) red deer males pretend indifference to their aggressive fellows, but as their battles go on the cheats creep up to a female and mate. Males of the ruff, a large wading bird, even become transvestites. They come in three genetically distinct forms. One has a black feathered neck, or ruff (*left*), in another the ruff is white (*right*), but the third group look just like females (*below*), and are accepted as such by the larger males, and can copulate with the females without being disturbed.

Cheats are everywhere. Flowering plants need a flying penis, an insect pollinator, to reproduce. The two are trapped in an endless contest. The insects want to be fat, lazy and promiscuous, but the plant prefers them to be hungry, active and faithful. Bees must visit a million flowers to make a pound of honey, but sometimes they get nothing, for bee orchids look like female bees and persuade males to visit them with no chance of sexual success.

Dishonesty goes further. Harmless flies bear yellow stripes that disguise them as wasps, while parasites sneak in with cellular cues that resemble those of the host. In the struggle for existence fraud is almost inevitable. If you want honesty, try physics instead.

Evolution by accident

Random events play a large part in evolution (as they do in physics). In 1929 a speedy French destroyer was sent from the African port of Dakar to deliver mail to Brazil. It had some unwelcome passengers: malaria-infected mosquitoes. Previous ships had taken so long to make the trip that the stowaways died en route, but the new vessel got there in record time, and some survived. Soon, a great malaria epidemic began.

There is now much less diversity in that parasite in Brazil than in Africa, for only a tiny sample had made it across the ocean. That single accident did more to change its identity than had thousands of years of natural selection in its native land.

Other diseases have been through even smaller bottlenecks. The agents of leprosy, smallpox and plague all descend from animal parasites. They are diverse in their original hosts but vary scarcely at all in their millions of human victims, as proof that they began with a single invasion. The agent of plague, for example, is almost identical across the globe; and the remains of bones from London's ancient epidemics bear almost the same genetic signature as do the plague bacilli of today. Perhaps a single bite of an unlucky human by an infected flea long ago led to a disease that has killed millions.

The inhabitants of remote islands have much the same history. Hawaii has its own unique plants, snails, insects and birds, but most have almost no DNA variation, evidence that each descends from a few ancestors who arrived long ago. One Act of God (if not God himself) can have dramatic effects on evolution.

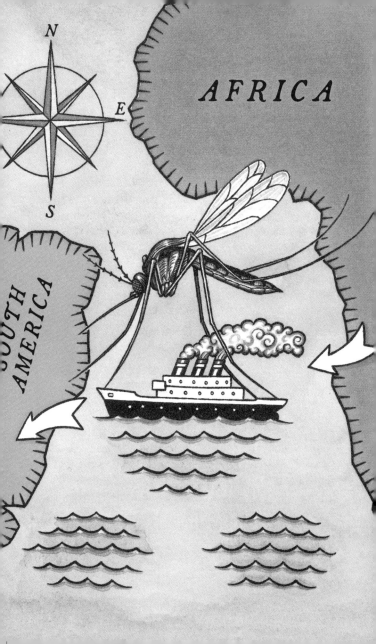

The republic of genes

The Origin of Species is mainly not about how species originate. What, indeed, is a species in the first place? It seems simple. Cats and dogs are unalike and cannot mate: they are different species. However, dogs and wolves also look unalike, but they can exchange genes. Are they different species, too?

Evolution is about change, and any attempt to fit all its products into rigid categories will not work. The same is true of language. English and Chinese are quite distinct, but although English, German, Danish and Dutch are dissimilar, their boundaries are not entirely clear. Across northern Europe, from east to west, there is a shift, manifest first as local accents and then as distinct languages. In western Holland, Friesian begins to sound remarkably like English; as the locals say, *Brea, bûter en griene tsiis is goed Ingelsk en goed Frysk* – Bread, butter and green cheese is good English and good Friesian.

Other locals also blur their identity. Crows (German, *Kräke*; Dutch, *kraii*; Danish, *krage*) come in two forms, the black carrion crow of the west and the partly grey hooded crow of the east. In a narrow zone they overlap and hybridize – but the hybrids do not spread. The same is true of mice (*Mäuse, muizen, mus*) and oaks (*Eichen, eiken, egetraer*). Each invaded from refuges in Greece and Spain as the Ice Age retreated, but all had evolved a separate identity in their long separation. In Central Europe, we can almost, but not quite, see the origin of species in action.

The fall of the wall

Mice are models for men, for their genetics is as well known as our own. Crosses have revealed the genes behind colour, behaviour and more, almost all of which involve many genes that interact with each other.

The two mouse species involved in the European hybrid zone diverged in isolation, thousands of years ago. As a result, some of those interactions break down when they spread and meet. Hybrid males in the zone are as a result less fertile, with fewer and slower sperm, than either parent. The eastern and western sperm factories do the same job, but in different ways. As in a production line set up with a mixture of machines from Trabant and Mercedes, quality has collapsed.

Elsewhere, matters have gone further. Within all species, cell division is controlled by genes that speed it up, balanced by others that slow it down. When this system goes wrong, cancer may result. Laboratory hybrids between two fish species, one with harmless black spots and the other without, lead to disaster, for in later generations the spots turn into lethal skin cancers and the hybrids die. Within each species cell division is controlled by a balance between accelerators and brakes. Put a Mercedes engine into a car with a Trabant's brakes and a crash is inevitable.

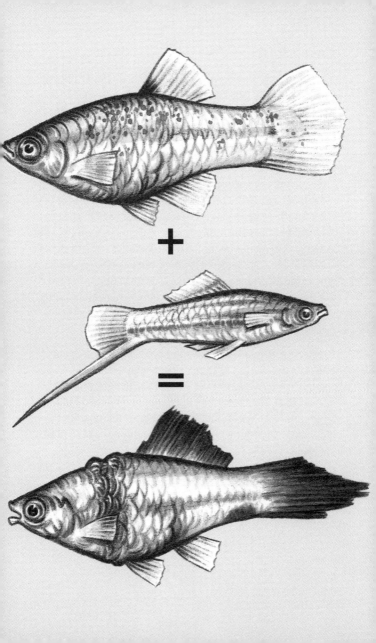

Cast adrift

The first Ladybird book was published in 1914. The
Atlantic is now two metres wider than it was then. The
continents skate around the globe as the liquid centre of
our planet bubbles up. Their arrangement today began
when the great land mass of Pangea broke up 200 million
years ago.

Continental drift solves many geographical puzzles.
Fossils of the same freshwater crocodile are found in
both South America and Africa, while those of an extinct
fern are in Australia, Antarctica, India, Africa and South
America. A family of modern earthworms – not fond of
salt water – is sprinkled across South America, Africa,
Madagascar, India, Australia and New Guinea. Indeed,
the genes suggest that elephants, aardvarks and dugongs,
diverse as they are, emerged from a shared ancestor when
Africa was an island.

Plants and animals have migrated not just through their
own efforts, but because their homelands drag them
along. Now, thanks to man's efforts with cars, ships
and aircraft, the world has in effect been reunited into a
single biological continent and many of its inhabitants –
elephants, aardvarks and dugongs included – have paid
the price.

PANGEA

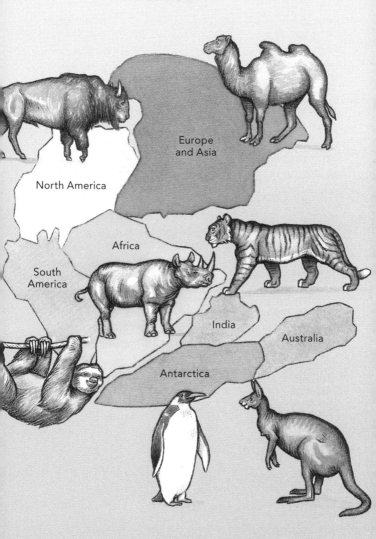

North America

Europe
and Asia

Africa

South
America

India

Australia

Antarctica

Dust to dust

The Battle of the Somme took place just over a century ago.
Now the site has a memorial marked with 72,000 names of
those British and Allied forces who 'had no known grave'.
The Allied cemeteries contain 150,000 crosses and the
Germans suffered even greater carnage.

A century on, many of the graves have no occupants, for
their bodies have returned to dust. The same is true of almost
all the 60 billion people who have lived and died since our
species emerged. The monuments of the First World War are
themselves crumbling, while the first pyramids, built to give
immortality to their occupants, have gone altogether.

Almost no bones – even those treated with reverence – rest
in peace. The fossil record is filled with great gaps as a
result, but what does survive tells a vivid tale of a long-lost
world. No more than one in a hundred of all the species
that have ever lived is alive today, and for most no relics
remain. Quite often whole communities have – like the
soldiers in the trenches – been eliminated. The dinosaurs
went 65 million years ago, perhaps because an asteroid
crashed into the world and darkened its skies for years. Four
earlier great disasters had killed off whole groups of marine
animals, but since 1914 we have seen what may be the
beginning of the greatest extinction of all, brought on, like
the First World War, by human stupidity.

The downy dinosaurs

Like many biologists, I started as a birdwatcher. Nowadays, students are more often drawn into the subject through a childhood fascination with dinosaurs. We share the same interest, for birds are just dinosaurs that flew away.

They did the job with feathers. We have hairs, but dinosaurs had scales, made of the same stuff. Each can be modified; hair into spines and, Chinese fossils show, scales into feathers.

The earliest feathered dinosaurs had only fluffy down, which may have kept them warm. In some lineages, those structures became larger, with a stout central shaft and fine branches on each side. Often they were on the tail or the head, perhaps as sexual displays. Some of those who bore them had no chance of leaving the ground, for they weighed a ton. In time, though, a few of the smaller kinds began to use feathered forearms to help pursue prey or slow a jump from a rock. As so often in evolution, birds did not emerge in a flying leap but slowly opened the door to a new way of life. Once they had found a way to exploit the air, safe from competition with other creatures with the same idea, evolution went into overdrive and generated nearly all the families of modern birds within a few million years.

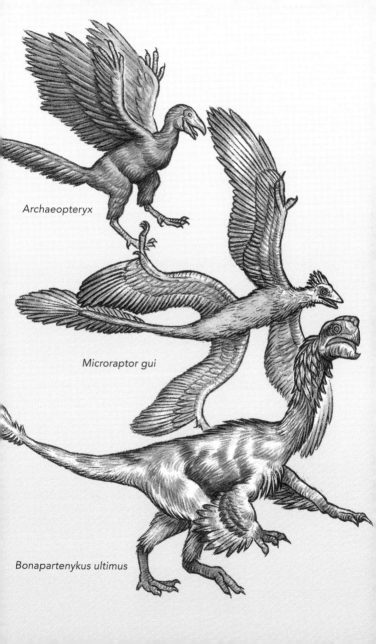

Archaeopteryx

Microraptor gui

Bonapartenykus ultimus

Roots and branches

Dinosaur relics, feathered or not, are rare indeed, but the information hidden in DNA means that every plant and animal is a living fossil. Darwin dissected many creatures in an attempt to uncover patterns of relatedness. Molecular evolution is no more than comparative anatomy plus a lot of money. Its chemical scalpels have reconstructed life's family tree.

It has an unexpected shape. Humans, amoebas, fruit-flies, bananas and elephants, all of which have cells with a central nucleus that contains the genes, sit on a minor branch, with mammals on an insignificant twig. Most of its limbs are devoted to bacteria and another group of single-celled beings called Archaea, in each of which the DNA floats free in the cell. Those two differ from each other, and from ourselves, in their chemical machinery. Archaea may be the most ancient life forms that themselves gave rise to plants and animals.

The pedigree of life has other surprises. Bacteria and Archaea often leak DNA between different kinds, so that their family trees begin to look like mangrove swamps and it is hard to see where one group ends and the next begins. Our own pedigree is more like a pine forest, but it is rather a shock to find that animals (ourselves included) are, in the great scheme of things, not very distinct from mushrooms. As *Homo sapiens* tries to find its place in the new taxonomic jungle, a certain humility is called for, although so far there has been little sign of it.

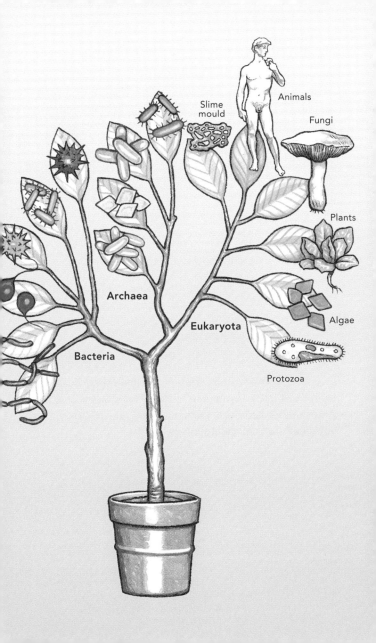

Animals

Slime
mould

Fungi

Plants

Archaea

Eukaryota

Algae

Bacteria

Protozoa

Hybrid history

On the evolutionary clock that began ticking at the origin of life 3,500 million years ago, land animals appeared at around 21.30, mammals at 23.30, early humans at two minutes to midnight and our own species some forty seconds before the chimes that rang in today.

The molecular fossils of our ancestors show that we live in unusual times. There is only one species of human around today, but for much of the past there were several. In northern Europe and Siberia, Neanderthals lived alongside us until about 40,000 years ago. They were heavily built creatures, adapted to cold. Denisovans – another kind of human – are known from just a few fragments in a Siberian cave but fossil DNA shows that they were distinct from both Neanderthals and ourselves.

That material tells an unexpected tale of the sexual habits of our ancestors, for paternity tests show that they mated with their relatives. As a result, about one part in fifty of modern European DNA (the gene for red hair included) is of Neanderthal origin. About the same proportion of Denisovan DNA is found in today's Melanesians and Aboriginal Australians. Outside Africa, at least, the ghosts of our vanished relatives live on.

Darwin in the kitchen

On its journey across the world, our species changed fast in the face of shifting challenges. Rickets, a disease caused by vitamin D deficiency, was once common in Britain. The vitamin is present in oily fish, but the poor could not afford such delicacies. It can also be made in the body when exposed to sunlight, but in smoky towns the sun never shone. The first emigrants from Africa also faced that problem, for dark skins do not let in much ultraviolet. In response, evolution favoured mutations that lighten the skin – and in the cloudy north and west produced blondes and redheads. The Khoi-San peoples of southern Africa are also lighter than those around the equator, proof that the rule applies as much in the southern hemisphere as in the northern.

Evolution did even more beneath the skin. Almost all Scots, but just half of Greeks, can drink milk when adult. Those who cannot feel ill when they try, for the enzyme needed to digest it has been switched off, as it is in almost all other mammals. In Scotland (but not in Greece) cattle herding began long ago, and those who inherited a new mutation that could digest the milk they produced had an immediate advantage. In the same way, peoples who eat lots of starchy food such as rice have more powerful enzymes to break down starch (and Europeans are, for reasons unknown, much better at breaking down alcohol than are many Chinese). Like sticklebacks, we evolved fast to cope with a new home.

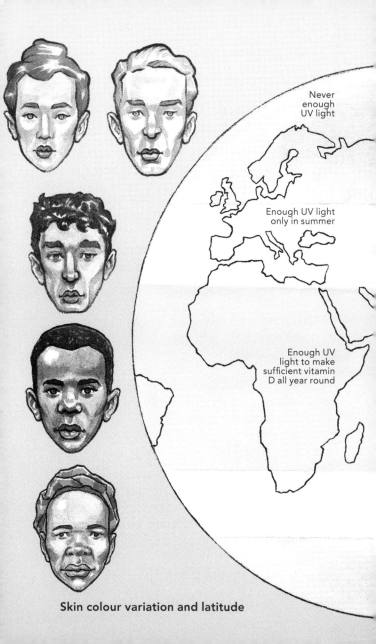

Never
enough
UV light

Enough UV light
only in summer

Enough UV
light to make
sufficient vitamin
D all year round

Skin colour variation and latitude

The endangered ape

Two hundred thousand wild chimpanzees are left, but *Homo sapiens*, their closest relative, is forty thousand times more abundant. Chimps may soon be extinct even as humankind fills the globe.

Once, the opposite was true. As the Brazilian malaria parasites show, small populations lose inherited variation. The island of Tristan da Cunha was settled in the nineteenth century by seven men and eight women. That tiny group of founders means that it now has less genetic diversity than does a British village of the same size. The same is true on other islands and the decrease is inversely proportional to the size of the founding population.

Fossils track our course out of Africa about 50,000 years ago, into Europe some 5,000 years later, and into the New World just 20,000 years before the present. There is a matching decline in diversity the further we go from our native continent, with no more than 60 per cent of the variation in the southern Andes compared with Ethiopia. The decrease can be used to estimate how many people made each step of the journey. The results are startling. They suggest that around 50 people were the ancestors of all those who made it out of Africa, 150 the ancestors of those who then moved on into Europe and around 100 the ancestors of all Native Americans.

However, Africans themselves are much less variable than are chimps – a hint that, even in its birthplace, *Homo sapiens* was once rare. Abundant as we might now be, for most of history we, not they, were the endangered species.

Local percentage of the genetic variation found in Africa
(defined as 100%)

| 100% |
| 90% |
| 80% |
| 70% |
| 60% |

→ Migration route

The future: feebler but smarter

The road leading from our shared ancestor with chimpanzees 7 million years ago is littered with discarded talents. We are as hirsute as chimps, but most of our hairs are downy. We have canine teeth, but they are unimpressive. Men have tiny testicles and, unlike chimps, no spines on the penis to hold a female in place. Even worse, we have tiny jaw muscles and feeble digestive enzymes. Humans cannot live on raw food alone and without an external stomach – a cooking pot or microwave – we would starve. In physical terms, *Homo sapiens* is a diminished chimpanzee.

When it comes to the grey matter, the opposite is true. We have brains four times bigger than those of chimps. Their talents have allowed us to fill the world. We have taken Africa with us in the form of clothes, lights and central heating. Most important, we have invented a new way to transmit information. Language does that far better than DNA and means that, uniquely, the arena of human evolution has moved from body to mind.

The world has been transformed. In Europe, life expectancy has gone up by six hours a day, every day, since 1945. Almost everyone lives long enough to have children, and most people have about the same number. As a result, natural selection, which depends on differences in survival and in fertility, has – at least for the time being and at least in rich countries – almost stopped. Jumbo jets mean the days of population bottlenecks have gone. In this new Pangea barriers are starting to break down. If stupidity does not destroy it, the future will see a triumph of the average as genes mix and merge. Fortunately, perhaps, none of us will be around to find out.

Further reading

Evolution is an enormous subject for it encompasses the whole of biology. There are various popular treatments (my own included):

Steve Jones *Almost Like a Whale: The Origin of Species Updated* (Black Swan, 2000) is an attempt to rewrite *The Origin of Species* using information from modern biology. It covers most of the topics discussed here.

Steve Jones *Y: The Descent of Men* (Abacus, 2002) does much the same for Darwin's second most famous book, *Sexual Selection and the Descent of Man*.

Jerry Coyne *Why Evolution is True* (Oxford Landmark Science, 2010) was initially aimed at a sceptical American audience, but gives a clear account of modern evolutionary theory, from fossils to DNA.

Richard Fortey *The Earth: An Intimate History* (Harper Perennial, 2005) deal with geology and the fossil record, a major element of Darwin's own argument.

Richard Dawkins and Yan Wong *The Ancestor's Tale: A Pilgrimage to the Dawn of Life* (Weidenfeld & Nicolson, 2005) is a survey of the whole world of life, its most obscure crannies included, from the perspective of an evolutionist.

And, of course, there is always Charles Darwin's *The Origin of Species, by Means of Natural Selection, or the Preservation of Favoured Races in the Struggle for Life*, first published in 1859 and now available in dozens of editions. Slightly hard work in places, but the foundation stone of modern biology. To see that Darwin really could write and to savour the sense of excitement of a young man gripped in a passion for science there is also *The Voyage of the Beagle* of 1839, in my view the greatest travel book ever written.